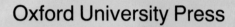

the Fly

Oxford University Press

Oxford Toronto Melbourne

A fly.

A frog.

A fly on a leaf.

A frog on a leaf.

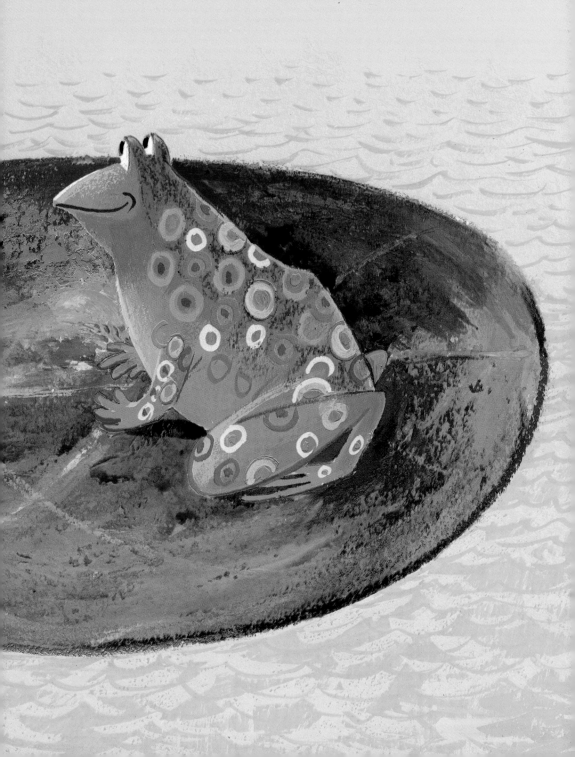

Frog catches fly.

Fly in the frog.

Fly tickles the frog.

Whoops . . .

. . . fly out of frog.

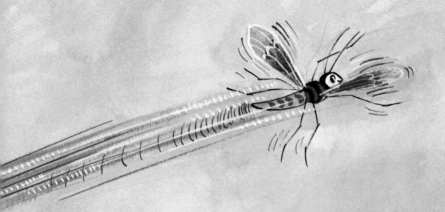

Oxford University Press, Great Clarendon Street, Oxford OX2 6DP

Oxford University Press – Education
198 Madison Avenue, New York, New York 10016

Oxford New York
Athens Auckland Bangkok Bogotá Buenos Aires
Calcutta Cape Town Chennai Dar es Salaam Delhi
Florence Hong Kong Istanbul Karachi Kuala Lumpur
Madrid Melbourne Mexico City Mumbai Nairobi Paris
São Paulo Singapore Taipei Tokyo Toronto Warsaw

and associated companies in
Berlin Ibadan

Oxford is a trade mark of Oxford University Press

© Leslie Wood 1985
First published 1985
Reprinted 1986, 1988, 1989, 1991, 1992, 1994, 1995, 1997, 1998, 1999 (twice)

This edition is also available in
Oxford Reading Tree Branch Library Stage 2 Pack **A**
ISBN 0 19 272144 5

British Library Cataloguing in Publication Data
Wood, Leslie
  The frog and the fly
  I. Title
  823'.914[J]   PZ10.3
  ISBN 0–19–272154–2
  USA ISBN 0–19–849015–1

Typeset by Oxford Publishing Services, Oxford
Printed in Hong Kong